In The Path

Glimpses into the Florida Treasure Coast's 2004 Hurricane Season.

Compiled and edited
by Scripps Treasure Coast Newspapers

In The Path

Dedicated to everyone on the Treasure Coast
affected by Hurricanes Frances and Jeanne,
and to the thousands of people who came from all across America
to help this community get back on its feet.

IN THE PATH

FORWARD

The two hurricanes that pounded the Treasure Coast in September left some lasting memories. One of my lasting memories will be the tireless, skillful work of our staff of photojournalists. They rode out the storms in strategic locations-- emergency operations centers, shelters and in one case a beach house.

They worked weekdays and weekends and 16-hour days were common. Eleven Scripps Treasure Coast Newspaper photojournalists took more than 20,000 images during and after the storms. Lloyd Young, picture editor for the papers, and Kelly Rogers, photographic assignment editor, made the final selections.

This book of photographs documents our community's September ordeal.

It also represents a benchmark and a year from now or years from now we can open it and be reminded of our community's resilience and ability to endure.

Tom Weber

Tom Weber

Publisher

IN THE PATH

HURRICANE FRANCES

image by ALEX BOERNER

Winnie Trudeau gets a little help from her neighbor to lower the storm shutters on her manufactured home in Barefoot Bay in Brevard County in preparation for the storm.

In The Path

Hurricane Frances

image by CHRIS WEATHERLY

Tom Gilowy, of Tampa, helps his mother, Karen McCumber, pack belongings from her home in Barefoot Bay. Gilowy visited his mother two weeks earlier, when Hurricane Charley hit the west coast of Florida. "I just got settled back into my home from Charley and now I'm moving stuff in for my mom," Gilowy said.

IN THE PATH

image by ALISON SIDLO

Britton Smith, left, and his father Carlton Smith show their team spirit by spray-painting on their hurricane shutters at Don Smith's Paint Store in Vero Beach.

In The Path

image by MATTHEW RATAJCZAK

Heeding the mandatory evacuation in effect for Martin County residents on Hutchinson Island and in mobile homes, Mike Lane, left, and his wife Mary, load up their pets, including one of their ferrets Snootcher, from their home at River Side Village Mobile Home Park on Indian River Drive in Jensen Beach. The Lanes made their way north like thousands of other South Florida residents as the hurricane bears down on the Treasure Coast.

IN THE PATH

image by ERIC HASERT

Louis Salem, 23, walks out from the house he shares with his brothers Simon, 25, and Raphael, 20, in Port St. Lucie. The brothers, who work as tree trimmers, tied down their home on Bayshore Boulevard to protect it from the coming hurricane. "We put in 6- to 8-foot anchors into the ground because we didn't want the hurricane to take down the house," Raphael Salem said.

In The Path

image by ALEX BOERNER

Winnie Trudeau applies duct tape to the outside of her manufactured home's front windows in Barefoot Bay in preparation for the approaching hurricane. The Brevard County resident, who went through Hurricane Irene in 1999, has grim expectations of the coming storm. "I've worked hard all my life and now I'm going to lose everything. It's depressing. I don't know why I bother taping the windows because I probably won't have a house when I come back," Trudeau said.

IN THE PATH

image by MOLLY BARTELS

Customers wait in line at the Steil Oil Citgo at 1031 U.S. 1 in Vero Beach to get their propane tanks filled in preparation for the hurricane. The line for propane remained steady from the time the station opened.

IN THE PATH

image by CHRIS WEATHERLY

Amos Bellefleur, an employee of East Coast Lumber in Vero Beach, takes a short break atop the last 15 sheets of 1/2-inch plywood remaining for purchase. The Vero Beach Police Department closed 11th Avenue for motorists waiting to enter the lumber yard to purchase 4x8 sheets of plywood. East Coast Lumber sold more than 5,000 sheets of plywood in two days.

IN THE PATH

image by MATTHEW RATAJCZAK

Martin County Sheriff's Deputy Jack Munsey informs beachgoers at Jensen Beach Park that they must get off Hutchinson Island because of a mandatory evacuation in effect as the hurricane approaches.

In The Path

image by ALISON SIDLO

Cody Cabrie, 10, of Fort Pierce videotapes the incoming squalls along the waterfront of the Indian River Lagoon in Fort Pierce. Cody and his mother and sisters were among the few Treasure Coast residents who ventured out of their shelters to watch the hurricane as it approached land.

image by DEBORAH SILVER

Tiffany Lenoir of Stuart stands in a line of people waiting to get into a shelter at Pinewood Elementary School in Port Salerno. Lenoir, who is seven months pregnant with her third child, said she was in the hospital the previous day because of anxiety. "Between this heat and not knowing, it's kind of hard to relax," Lenoir said.

In The Path

image by JUAN DALE BROWN

With hundreds of cots set up at the St. Lucie County Civic Center, one of the county's special-needs shelters, Dawn Bryan settles in with her children Julia Hosking, 4, Ian Hogan, 7, and Symone Hogan, 10. Bryan said she thought it would be safer to wait out the hurricane at the Fort Pierce shelter.

In The Path

image by SARAH GRILE

George Scheussler, who lives in congregate living in the Florida Baptist Retirement Center, sleeps soundly inside the First Baptist Church in Vero Beach during the hurricane. He and other center residents stayed at the church for four days, including an extra day spent there because of flooding in parts of the Vero Beach retirement center's buildings.

IN THE PATH

image by ALEX BOERNER

Louise Ebanks, of Vero Beach, stares through the shelter's rain-covered doors into the courtyard of Oslo Middle School, after the storm shutters were opened and people were allowed to leave the Vero Beach school's protection in the afternoon.

IN THE PATH

image by ALEX BOERNER

From left, Elisa Green, Yashira Estrella, 12, Cor'nasia Potter, 5, and Dottie Powers, all of Vero Beach, participate in a morning prayer circle at Oslo Middle School in Vero Beach, where hundreds of people were taking shelter ahead of the hurricane. "We started this praying yesterday, and you can see this hurricane is now all confused," said Powers, commenting on the fact that the hurricane dropped from a Category 4 to a Category 3 two days before making landfall. It came ashore as a Category 2 storm.

IN THE PATH

image by ALEX BOERNER

Alice Szombaty, left, and her longtime friend and neighbor Sally Logan play a game of Kings in the Corner at the special-needs shelter at the Port St. Lucie Community Center. About 200 Port St. Lucie residents rode out the storm in the shelter.

IN THE PATH

image by SARAH GRILE

Pearl Ragan, who lives in the Florida Baptist Retirement Center's assisted-living unit, finishes her breakfast as the hurricane's winds come to an end. Ragan took shelter with other retirement center residents at First Baptist Church in downtown Vero Beach.

IN THE PATH

image by SARAH GRILE

Counterclockwise from right, Lori Ross, social services director of the Florida Baptist Retirement Center; Jean Chasteen, activities assistant at the retirement center; and Leigh Ann McDaniel, daughter of retirement center administrator Tommy McDaniel, try to make the best of the hours inside the First Baptist Church in Vero Beach as the hurricane rages outside. The storm's strong winds could be heard inside the building. Residents from the Vero Beach retirement center also took shelter at the church.

IN THE PATH

image by DEBORAH SILVER

Martin County Fire-Rescue Lt. Johnny Recca tests the strength of the wind as it reaches 50 mph in Jensen Beach. Fire-Rescue officials were busy placing crews together and preparing for post-storm issues.

IN THE PATH

image by MOLLY BARTELS

Vero Beach Police cruisers take refuge in a Vero Beach car wash's bays.

image by MOLLY BARTELS

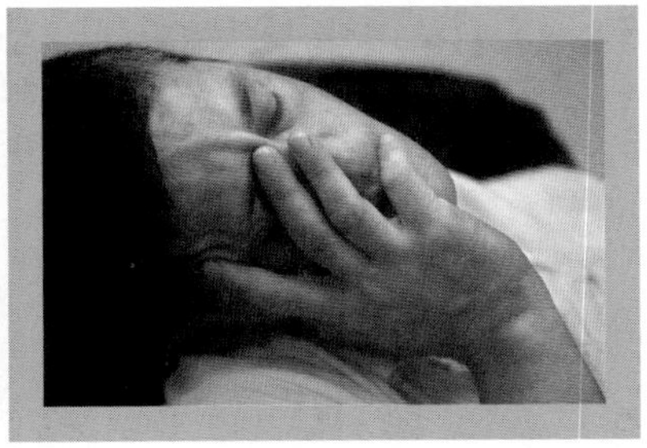

After several days with little sleep, Joe Flescher, spokesman for the Indian River County Sheriff's Office, catches a nap at the Emergency Operations Center in the county administration building in Vero Beach.

IN THE PATH

image by DEBORAH SILVER

Martin County firefighter Clint Bradley, 29, stretches out as he talks to his parents in Ocala at the county's emergency operations center in Jensen Beach. "I was just trying to pick a quiet spot and talk to my family," Bradley said.

"I was just trying to pick a quiet spot and talk to my family."

IN THE PATH

image by DEBORAH SILVER

Martin County firefighter Cheryl Recca, center, watches the massive storm's nighttime progress with others gathered around her in Jensen Beach as the hurricane makes its way toward the coastline.

IN THE PATH

image by ALISON SIDLO

A sailboat at the Fort Pierce Marina begins to take hits as the hurricane's outer bands enter the area.

IN THE PATH

image by JUAN DALE BROWN

In blowing afternoon rain, members of the Florida National Guard Taskforce 53 SIB establish their headquarters at the St. Lucie County Fairgrounds. About 2,000 National Guard members assisted local government agencies in more than 11 counties.

IN THE PATH

image by ALEX BOERNER

Nathan Lazinsky, of Port St. Lucie, who was weathering the hurricane with his friend, Dory Klock, of Vero Beach, removes a sunken piece of siding while working to clear out a drain near the entrance of Vero Beach's Laurelwood development, which sustained significant flooding.

IN THE PATH

image by DEBORAH SILVER

Bobby Puckett, 39, and his sheltie, Benson, view the damage in his neighborhood at Ocean Breeze Park in Martin County. "It's pretty bad," said Puckett. "We're just getting out to see the damage."

Cecilia Cruz, left, looks over the damage from her home at River Side Village Mobile Home Park in Jensen Beach. "This was my little castle," said Cruz, who's lived in the park for five years. Cruz, 56, found the house marked by emergency personnel as uninhabitable. She has no insurance. Cruz stayed with Rose Ann Ricciardi, of Port St. Lucie, during the storm.

image by DEBORAH SILVER

"This was my little castle."

IN THE PATH

image by ALEX BOERNER

Rebecca Davis looks up at the damage done to the ceiling of the home she shares with her boyfriend, Mark Master, in the Laurelwood development in Vero Beach. The ceiling in many of the home's rooms collapsed after the roof collected too much rainwater from the hurricane.

In The Path

image by ALISON SIDLO

Max Delgado carries a mop past the ruins of the waterfront dining room at Chuck's Seafood Restaurant along the Fort Pierce Inlet on Seaway Drive. Despite the damage, employees were working to re-open the restaurant.

IN THE PATH

image by MATTHEW RATAJCZAK

Ricky Malando, 16, of Stuart, hauls away a bag of debris at his family's Stuart business, Rapco Auto Electric, after the hurricane destroyed the building during its rampage through the Treasure Coast. The family thinks a tornado did most of the damage.

In The Path

image by MATTHEW RATAJCZAK

A variety of boats rest in a heap at the mangled Fort Pierce Marina after the hurricane blew through the Treasure Coast.

In The Path

image by MATTHEW RATAJCZAK

The hurricane's sustained winds blew the roof off a hangar at the St. Lucie County International Airport and damaged many other buildings and planes on the property.

IN THE PATH

image by MOLLY BARTELS

After waiting several hours for provisions trucks to arrive, Eric Hodges, front, and Reginald Bien-Aime, back, of the Florida Army National Guard, take naps in the parking lot of the Indian River Mall in Vero Beach. The trucks did not arrive because of severe weather in North Florida.

image by MOLLY BARTELS

Florida Army National Guard Lt. Marc Ramello manages a quick shave in the parking lot of the Indian River Mall in Vero Beach. Ramello's Miami-based company was in the city to aid local law enforcement.

IN THE PATH

image by MATTHEW RATAJCZAK

A Florida National Guard soldier speaks with a motorist waiting to get back onto Hutchinson Island in Martin County.

In The Path

Zelda Jarbis, 84, of Jensen Beach, talks about her circumstances to Gov. Jeb Bush, who was touring the Treasure Coast after the hurricane. Jarbis and her husband, Daniel, 82, have been at the special-needs shelter at Challenger School in Port Salerno for five days.

image by DEBORAH SILVER

Gov. Jeb Bush makes a morning visit to the Brevard County Emergency Operations Center in Rockledge to discuss procedures for dealing with the approaching hurricane.

image by MOLLY BARTELS

IN THE PATH

image by MOLLY BARTELS

First Lady Laura Bush poses with a crowd of volunteers at a hurricane relief distribution center at the Indian River Mall in Vero Beach. Bush helped members of the National Guard load water and ice into vehicles.

In The Path

image by MATTHEW RATAJCZAK

At the Evans Crary Bridge, Martin County Sheriff's Office Deputy Joe Martin, left, explains to Sharon Weston, who lives in a Hutchinson Island condominium, why she and other island residents are not allowed back to their homes or businesses. The Sheriff's Office and National Guard troops were stationed at the bridge in Jensen Beach to redirect traffic and explain that the Department of Transportation must inspect the bridge before it can reopen. Residents were allowed back later in the day.

In The Path

image by JASON NUTTLE

Elizabeth Johnson, 65, a resident of Ocean Breeze Park, voices her concerns about the lack of power and utilities in the Martin County trailer park community. Johnson is taking care of a 91-year-old neighbor, Juanita Deykes, whose house was destroyed. Johnson and the rest of the 100 residents who attended the meeting were told it could be a couple of weeks until power is restored. In the meantime, the Federal Emergency Management Agency and Martin County have offered assistance until power is restored.

IN THE PATH

image by DEBORAH SILVER

A worker walks off with a breakfast plate as others line up for the same before heading out to work at the Florida Power & Light Co. staging area at Witham Field in Stuart. The utility workers spend the night at hotels — some as far away as Boca Raton— and return to Witham in the morning to receive instructions for the day.

image by DEBORAH SILVER

Bumper stickers on the back of a truck at Witham Field refer to the two biggest disasters to strike Florida so far. It is also a source of income for the utility workers who call themselves storm chasers. One power worker said he follows the hurricanes because that's where the money is.

IN THE PATH

image by JUAN DALE BROWN

Crewmen from NG Gilbert, an Indiana-based utility contractor, work on power lines along State Road A1A on North Hutchinson Island. It took the crews three days to get to the Treasure Coast, said wireman Rex Rhea, of Eminence, Ind.

IN THE PATH

image by SARAH GRILE

Shandra Baber, who lives off 19th Street in Vero Beach, has no explanations for why those living in her neighborhood have not received power and says she is "fed up with it." Baber decided to take her homemade sign to U.S. 1 to get the attention of passing motorists.

In The Path

image by ALEX BOERNER

Nell Obaker, of Vero Beach, watches from her porch as workers from Buffalo, N.Y., and Jackson, Miss., working for Florida Power & Light Co., repair downed poles in the back yard of her home on 35th Avenue.

IN THE PATH

A mobile home in Port Salerno provides electricity to a neighbor's powerless home via an extension cord stretched across the street.

image by MATTHEW RATAJCZAK

In The Path

image by MOLLY BARTELS

The lifeguard station and boardwalk at Wabasso Beach in Indian River County sustained major damage.

IN THE PATH

image by ALISON SIDLO

"I survived Hurricane Frances" T-shirts hang under a pop-up tent in the parking lot of a damaged gas station along U.S. 1 in Fort Pierce. Kris Maerz and Brian Wills (not shown), were among several vendors who turned a profit selling Hurricane Frances T-shirts and memorabilia from roadside stands throughout the Treasure Coast.

IN THE PATH

image by JASON NUTTLE

Brandon Miller helps Tom Arant (not shown) pick up debris around Arant's Garden Center off U.S. 1 north of Stuart in preparation for Hurricane Jeanne. The artist who occupies the building, Michael Brandes, painted a Hurricane Frances mural on the side of the business telling everyone driving on U.S. 1 to "Hang On" during the hurricanes.

IN THE PATH

image by ALEX BOERNER

Francis Hansen's Camaro Z-28 was lifted into the air at his Wabasso home when the tree it was sitting under was uprooted by the hurricane as it came through Indian River County.

IN THE PATH

image by DEBORAH SILVER

A derelict boat sits alongside a washed-out portion of Indian River Drive in Jensen Beach. The approximately 20-mile stretch that extends from Sewall's Point to Fort Pierce along the Indian River Lagoon received significant damage after 105-mph winds pounded the Treasure Coast for hours.

IN THE PATH

image by CHRIS WEATHERLY

Indian River County Environmental Health Specialist Julianne Renk looks through debris on the floor of Canaanland Missionary Baptist Church in Wabasso. The Indian River County church building lost a large part of its roof and sustained major damage to the front entrance.

In The Path

image by CHRIS WEATHERLY

A depiction of the Crucifixion lays among the debris at Canaanland Missionary Baptist Church in Wabasso in Indian River County. The building sustained major damage, including the loss of a section of the roof.

In The Path

image by CHRIS WEATHERLY

Ray Luder holds up a piece of lattice in the shape of a cross that he found standing straight up in a pile of debris when he returned to his manufactured home in Fairlane Harbor in Vero Beach.

image by MOLLY BARTELS

The steeple at Forest Park Baptist Church in Vero Beach ended up stuck in the roof after the hurricane hit the Treasure Coast.

IN THE PATH

image by ALEX BOERNER

Christ By The Sea's children's and youth pastor, Susan Brogan, celebrates after getting a window open that had been stuck shut. The Vero Beach church, which was without air conditioning, still conducted a Sunday morning service. Brogan was trying to make it as comfortable as possible for those who showed up.

IN THE PATH

image by ALISON SIDLO

Seth Soule, standing bottom, and David Izabal, both of Big Tuna Salvage, hook up equipment to hoist damaged boats onto a barge while Andy Swartz waits at the Fort Pierce Marina. Salvage crews are working around the clock pulling damaged boats out of the Indian River Lagoon.

IN THE PATH

image by DEBORAH SILVER

Marzena Rogers clears debris from around a gutter in downtown Stuart. Rogers, a saleswoman at Gore Creek Outfitters, said the store's awning was ripped down during the hurricane, but there was virtually no other damage. "We'll be ready in a day or two," she said.

"We'll be ready in a day or two."

In The Path

image by MATTHEW RATAJCZAK

Rick Binder, a boat captain with the Environmental Studies Center in Jensen Beach, removes pieces of siding that blew off surrounding mobile homes into the Indian River Lagoon.

IN THE PATH

image by SARAH GRILE

Kevin Hamblin, who lives on 4th Avenue in the Rockridge neighborhood in Vero Beach, cleans up the insulation that fell from his ceiling. Hamblin, who shares the rental house with his fiancee, Heather Mitchell, and 15-month-old baby, Mitchell Hamblin, said the Federal Emergency Management Agency inspector who looked over the house said "it was one of the worst standing structures he had seen" while in Vero Beach. Hamblin and his family, who have lived in the house for more than a year, had to move out because the house was condemned.

IN THE PATH

image by MOLLY BARTELS

Darwyn Rushing of Triple C Enterprises of Palm Bay reduces downed trees into haulable pieces on State Road 60 near 71st Avenue west of Vero Beach.

In The Path

image by JUAN DALE BROWN

Mounds of vegetative debris that will be turned into wood chips litter the St. Lucie County Fairgrounds at Midway and Okeechobee roads. Debris continues to be collected from roadsides, where residents cleaning up their property pile it up.

IN THE PATH

image by DEBORAH SILVER

Hiller Masker, left, grabs a backpack from his wife, Chris, as the two gather mud-covered personal items from their 32-foot Westsail boat that crashed ashore during the hurricane. The boat had been in the moorings at the Southpoint Anchorage in Stuart. "We're basically homeless because that's been our home for the past 25 years," said Hiller, 81, about the boat they named the Terra.

"We're basically homeless because that's been our home for the past 25 years."

IN THE PATH

image by DEBORAH SILVER

"It's all mud inside," said Chris Masker, referring to the galley of her sailboat as she picks out a mud-covered photo album and flips through it to decide whether it's too damaged to keep.

IN THE PATH

image by ERIC HASERT

Darrill Rimes, 28, of Fort Pierce, relaxes while making a morning phone call to a friend from a functional pay phone on U.S. 1 in Fort Pierce that was flattened by the hurricane. Rimes said it was the only working phone he could find.

IN THE PATH

image by DEBORAH SILVER

Paul Hefflefinger, of Sarasota, cools off in front of an industrial-size fan at Crossroads Community Church in Palm City. Hefflefinger, 78, is a volunteer with the Southern Baptist Convention Disaster Relief/Florida and has spent the past few days helping prepare thousands of meals for the American Red Cross Martin County chapter.

IN THE PATH

image by ALEX BOERNER

Jenny Dixon, 22, who has worked at the Vero Beach Humane Society for 10 years, untangles herself while walking five dogs at night. People working at the shelter have been continuously caring for the animals. Because many dogs are being kept in crates, rather than kennels, walking all of them is an ongoing process that often runs late into the night.

In The Path

image by ALISON SIDLO

Wyn Burns removes a rooster out of high water at the Animal Safe Place in Fort Pierce. Burns cares for injured and sick wildlife on her property, but is struggling to keep her animals dry after heavy rains from the remnants of Hurricane Ivan caused the Eleven Mile Creek and other bodies of water to jump their banks.

In The Path

image by DEBORAH SILVER

"We've waited in line for more than an hour," said Matt King, of Jensen Beach, as he takes a break from pushing his friend's car to save gas while waiting to fill up at the Shell station at Colorado Avenue and U.S. 1 in Stuart. The line of cars on Colorado stretched all the way to Confusion Corner with more than an hour wait. Gas station attendant Michal Figueredo said the station put a limit on gas purchases for a couple of hours before confirming a gas truck was on its way.

"We've waited in line for more than an hour."

In The Path

image by ALISON SIDLO

Devin Steele, 4, left, and Mark Jemison, 6, both of Fort Pierce, sit and wait on a cooler while Devin's mother stands in line for water and ice at Lawnwood Stadium in Fort Pierce. More than 5,000 people waited in line for supplies.

IN THE PATH

image by MOLLY BARTELS

Sam Pyles rests on an empty gas can outside Mark's Mobil service station in Vero Beach. Several hundred people lined up in cars and on foot around the station to get gas for vehicles and generators. The hurricane left Vero Beach without power, water and telephone capabilities.

IN THE PATH

image by JASON NUTTLE

Derek Wright, 24, of Port Salerno, pushes his car into a gas station off U.S. 1 in Stuart. Inside it are his girlfriend, Shanika Bedford, and their 1-year-old son, Deshawn. Wright had to push the car after the family ran out of gas while waiting in line for more than an hour to fill up.

In The Path

image by ALEX BOERNER

Sally Constantini, of Micco, takes a drink of water and tries to stay cool while waiting in line at the emergency disaster relief station near Barefoot Bay. "I was hoping I could get a tarp or something. If my roof starts leaking again, I'm in trouble," said Constantini, whose home sustained roof damage. People began showing up before 7 a.m. for the relief. Brevard County Sheriff's Office deputies and Red Cross volunteers handed out water and umbrellas to help people keep cool until relief supplies arrived about 12:30 p.m.

"It's gold."

image by ALEX BOERNER

"It's gold," said Christina Lape, of Barefoot Bay, who got a handfull of ice before the trucks containing bags of ice showed up at the emergency disaster relief site in Barefoot Bay in Brevard County. People who waited in line received five bags of ice per family.

IN THE PATH

image by ERIC HASERT

Bob Tolbert, center, boyfriend of store manager Julie Balma, keeps track of customers from the doorway of the Go Stop Express convenience store at Port St. Lucie Boulevard and Cameo Drive in Port St. Lucie. Tolbert and Balma went to the store to check for damage, but wound up opening for cash customers needing supplies for about three hours despite being without power.

In The Path

image by MOLLY BARTELS

Judy Woods, center, scoops vegetables into a take-out container inside a mobile feeding canteen run by the Salvation Army. Woods, who has worked in kitchens most of her life, was volunteering her time and services.

IN THE PATH

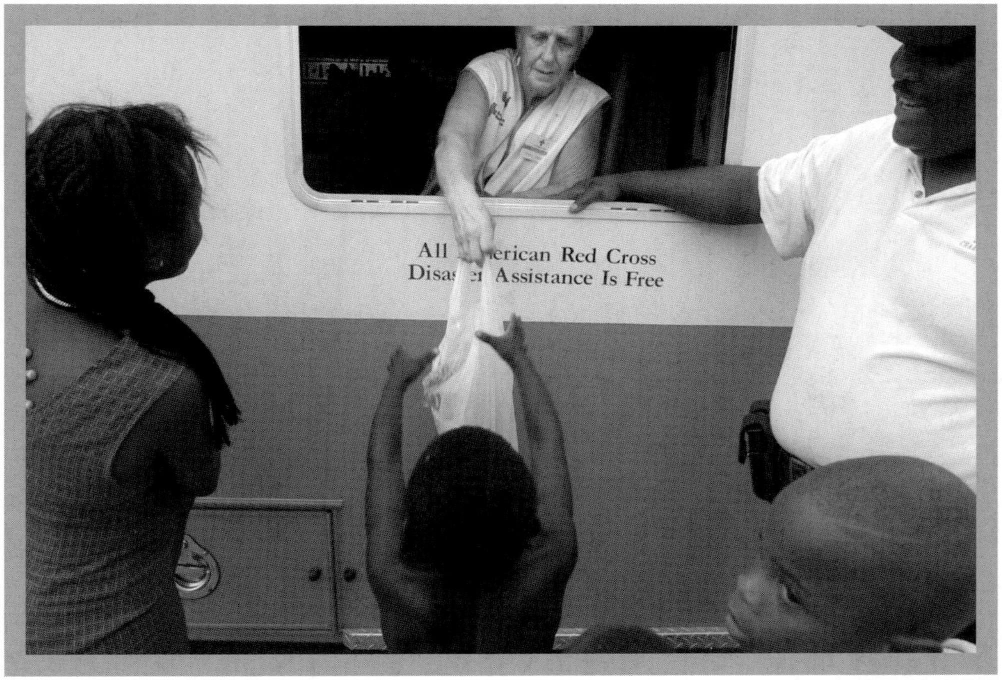

image by CHRIS WEATHERLY

Tootie Johnson, center, a Red Cross volunteer from Michigan, hands out Meals Ready-to-Eat to residents of Gifford in Indian River County. Johnson has volunteered for 48 years with the national relief agency and has been on hundreds of relief missions.

IN THE PATH

image by MOLLY BARTELS

Tilbert Ray Brickhouse, a volunteer from Tennessee, does a load of laundry in the back of a mobile laundry unit set up in the parking lot of the First Church of God in Vero Beach. The facility has two washers and dryers and six showers, three each for men and women.

IN THE PATH

image by MATTHEW RATAJCZAK

New Mexico Medical Disaster Team pharmacist Ben Kesner, left, leads Julia Aloi from the team's tented area back into Martin Memorial Hospital South in Port Salerno. The volunteer medical team has treated hundreds of patients as it helps ease the workload at the hospital's overflowing emergency room.

IN THE PATH

image by ALEX BOERNER

Ed Foxwell, of Micco, gets a handful of ice from Brevard County Sheriff's Office Deputy Crystal Petty at the emergency disaster relief distribution center in Barefoot Bay.

IN THE PATH

Susan Romano, left, a volunteer with the American Red Cross Martin County chapter, passes bottled water to Ashley Mason-Ovens, 11, as the emergency relief vehicle drives throughout Palm City. "It's cool that they give out food," Ashley said about the nonprofit organization.

image by DEBORAH SILVER

"It's cool that they give out food,"

IN THE PATH

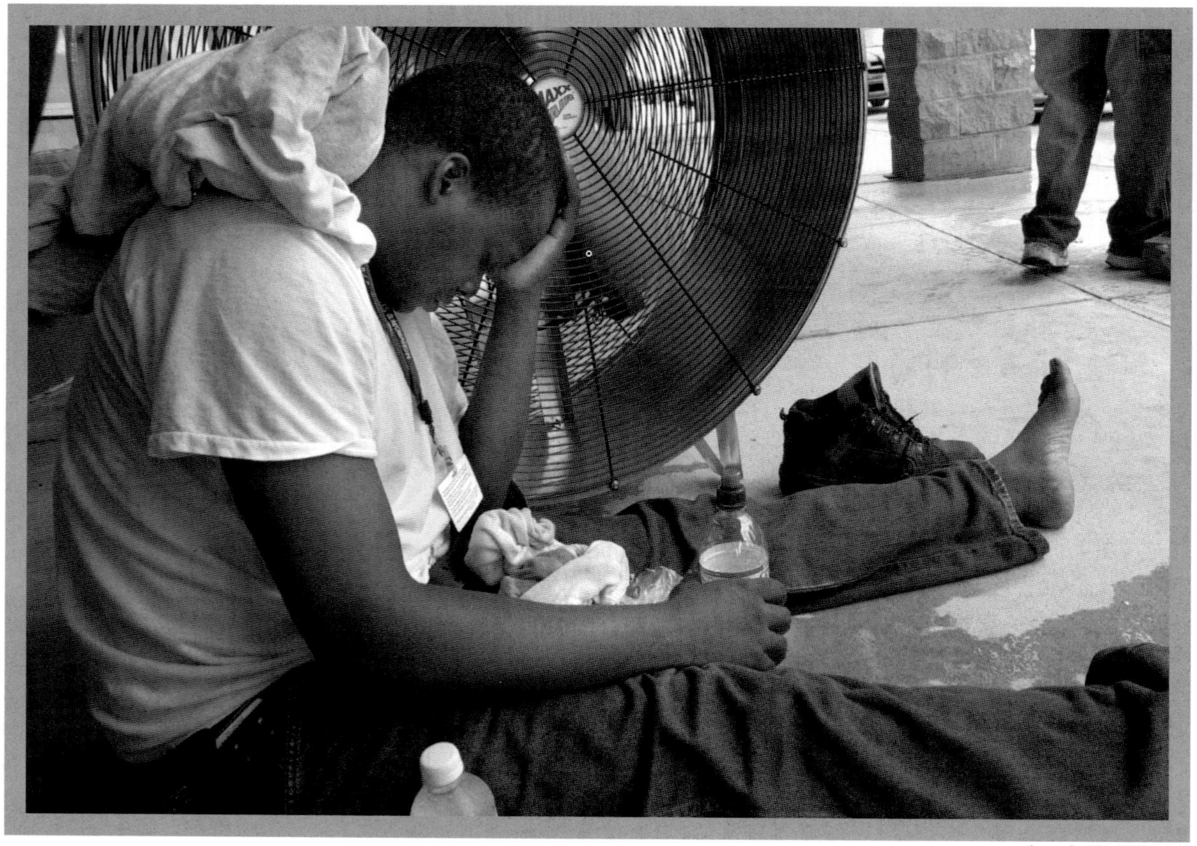

image by ERIC HASERT

Gregory Martin, of the Port St. Lucie Parks and Recreation Department, is overcome with exhaustion while helping give out supplies to hurricane victims at the Federal Emergency Management Agency distribution center at City Center in Port St. Lucie.

IN THE PATH

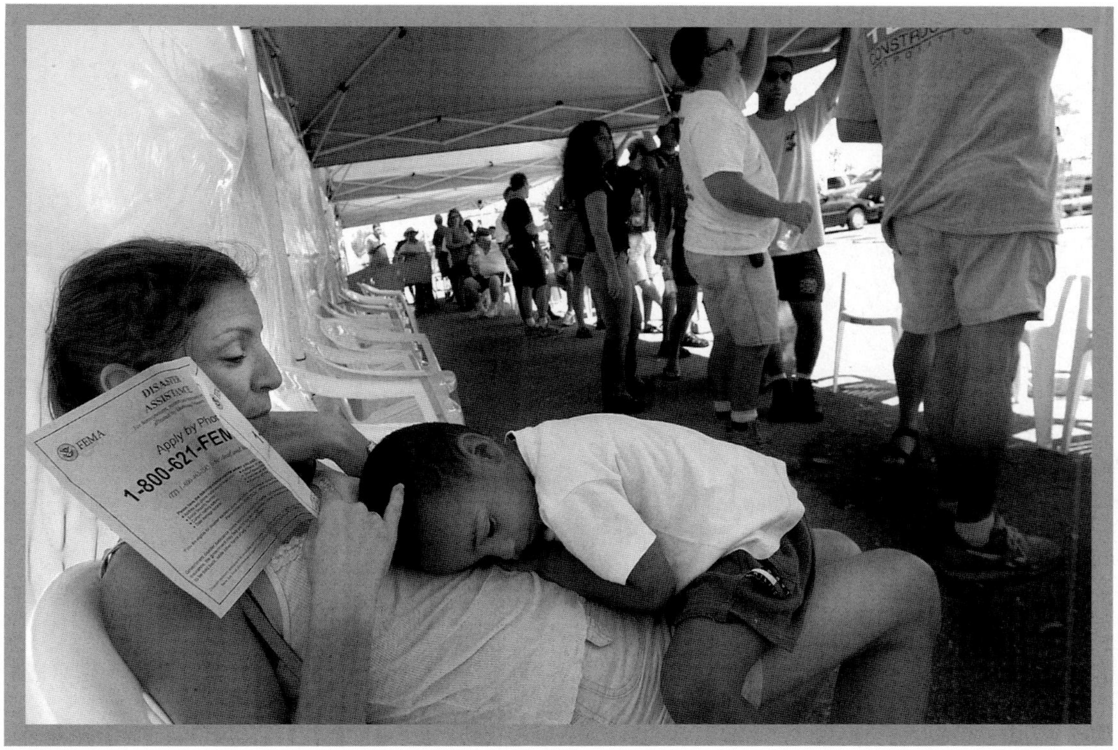

image by DEBORAH SILVER

Frankie Gardner, left, of Port Salerno comforts her son, Justin, 2, who grew tired of waiting in line for a generator at The Home Depot in Stuart. The wait for the coveted item was going on close to an hour. While Gardner was there, a shipment of 200 generators came in. "We've been looking for six or seven hours," said Gardner. "If that hurricane comes, then we'll be happy to have a generator."

IN THE PATH

image by ALEX BOERNER

Matt Guthrie, 23, makes his way through the fallen trees and floodwaters on a road near the Wabasso Causeway that leads to the Indian River Lagoon in Indian River County. Matt and his brother, Jesse Guthrie, 20, who are not residents of the barrier island, were going to try to paddle across the lagoon to get to the ocean to surf. "We're not trying to loot or anything, we just want to catch some waves. It's not every day you get to catch hurricane waves," Matt Guthrie said.

IN THE PATH

image by SARAH GRILE

Morgan Spencer, 11, jumps onto Laura Seymour's front porch in Rockridge in Indian River County. Seymour, who was busy moving her valuables out of her home, came home after the hurricane to find that her house was severely flooded and most of her family's belongings were damaged.

IN THE PATH

image by ALISON SIDLO

A Fort Pierce street sign pushed over by high winds lays in a yard near a boarded up house. Traffic lights and street signs throughout the region were down, making driving difficult.

IN THE PATH

HURRICANE JEANNE

image by CHRIS WEATHERLY

As the hurricane approaches the Treasure Coast, Donathan Williams looks out over Conn Beach in Vero Beach. Williams surfed earlier that morning, but cut the adventure short because of strong southern currents.

IN THE PATH

image by ERIC HASERT

Residents spray-painted their thoughts on boarded-up homes in Port St. Lucie.

IN THE PATH

image by MOLLY BARTELS

After waiting in line to fill his family's propane tank, Zachary Custar, 9, entertains himself while his mother, Karen, right, fills up extra gas cans at the Steil Oil Citgo at 1031 U.S. 1 in Vero Beach. Lines for gas and propane have been steady since the gas station opened.

IN THE PATH

image by JASON NUTTLE

Greg Marshall helps direct a 29-foot Boston Whaler into the Sundance Marine indoor storage in Jensen Beach as boat yard employees prepare for the hurricane. Sundance Marine's indoor shelter sustained significant damage from Hurricane Frances, but it's expected to be strong enough to shelter their customers' boats during Hurricane Jeanne.

IN THE PATH

image by MATTHEW RATAJCZAK

Eileen Eiler, 80, of Jensen Beach, puts up her hurricane shutters for the second time in a month. Eiler sustained some damage from Hurricane Frances and decided not to wait to the last minute to batten down for Hurricane Jeanne.

IN THE PATH

image by CHRIS WEATHERLY

Indian River County Emergency Management Coordinator Nate McCollum tapes a 1 p.m. weather broadcast from CCTV Studio at Vero Beach High School updating residents on the hurricane's projected path.

IN THE PATH

image by ALEX BOERNER

As the storm begins to move out of the area, the flooding to the barrier island in Vero Beach becomes evident. Ocean Drive, looking north toward the Village Spires, was under water.

IN THE PATH

image by ALEX BOERNER

Cameron Schneider,15, who stayed in his apartment with his father on the barrier island in Vero Beach, finds the family cat, B.C., hiding under the sink late at night during the hurricane.

IN THE PATH

image by ALEX BOERNER

Cameron Schneider, 15, climbs the lifeguard tower to get a better feel for the winds as bands from the hurricane hit near South Beach in Vero Beach.

IN THE PATH

image by ALEX BOERNER

From left, Guillermo Gonzalez and Oscar Hernandez, both of Miami, wave to neighbor Bob Schneider after emerging from their hotel room at South Beach Place on the barrier island in Vero Beach, where they rode out the hurricane. The two men were in Vero Beach to make repairs on the hotel, and decided to stay rather than fight the traffic getting back to Miami.

IN THE PATH

image by CHRIS WEATHERLY

From left, Carly Parker, 12, Julia Wilson, 14, and Riley Parker, 9, all of Vero Beach, enjoy the hurricane's wind gusts as they play outside on a deserted 6th Avenue in Vero Beach before weather conditions deteriorated to tropical storm-force winds.

IN THE PATH

image by ALEX BOERNER

After trying to dig open the door of a maintenance shed, Cameron Schneider, 15, watches as Dominic Pisano, 23, squeezes his way out of an opening near the top before the hurricane made landfall on South Beach in Vero Beach. "I thought I could wait it out," said Pisano, who realized it was a bad idea as water from the storm swell continued to flow into the small beach structure.

IN THE PATH

image by MOLLY BARTELS

Chris Gales surveys the damage at Palm Paradise mobile home park in Vero Beach the day after the hurricane hit the Treasure Coast. Gales stayed with his father, Oddy Gales, who would not leave his mobile home.

IN THE PATH

image by SARAH GRILE

Elliot Harrington, left, and Ed Lavandera, right, help Billy Dodds after he falls into the floodwaters at Fairlane Harbor in Vero Beach. Harrington and Dodds came to check the damage on their home and Lavandera, who works for CNN, traveled to the area to cover the hurricane.

IN THE PATH

image by CHRIS WEATHERLY

From left, Tim Poe, Candace Fersch, 14, and Joshua Thompson, 15, use a small boat to cross 18th Avenue in the Vero Beach Highlands after the hurricane's rains flooded the road.

In The Path

image by MOLLY BARTELS

Evelyn Auty is escorted out of Fairlane Harbor to her car by her neighbor, Jim Gearhardt. The manufactured home park in Vero Beach survived Hurricane Frances with only minor damage, but sustained more severe losses in Hurricane Jeanne.

IN THE PATH

image by JASON NUTTLE

A large section of MacArthur Boulevard on Hutchinson Island in Stuart was washed out by the hurricane, leaving the town of Sailfish Point disconnected from the mainland.

In The Path

image by JASON NUTTLE

An aerial view of Causeway Mobile Home Park in Fort Pierce shows the devastation caused by the hurricane.

IN THE PATH

image by JASON NUTTLE

A motorist walks away from his flooded car and back into the Spanish Lakes Country Club north of Fort Pierce.

IN THE PATH

image by JASON NUTTLE

Hundreds of people stand in line at a downtown Vero Beach business on 14th Avenue waiting to purchase generators.

image by JASON NUTTLE

A U.S. Marine pauses before opening the doors to Marine One before President George W. Bush and his brother, Florida Gov. Jeb Bush, toured the relief efforts from Hurricane Jeanne.

image by JASON NUTTLE

U.S. President George W. Bush waves as he and his brother, Florida Gov. Jeb Bush, exit Marine One as they arrive in Stuart to visit Hurricane Jeanne victims and the relief effort Sept. 30.

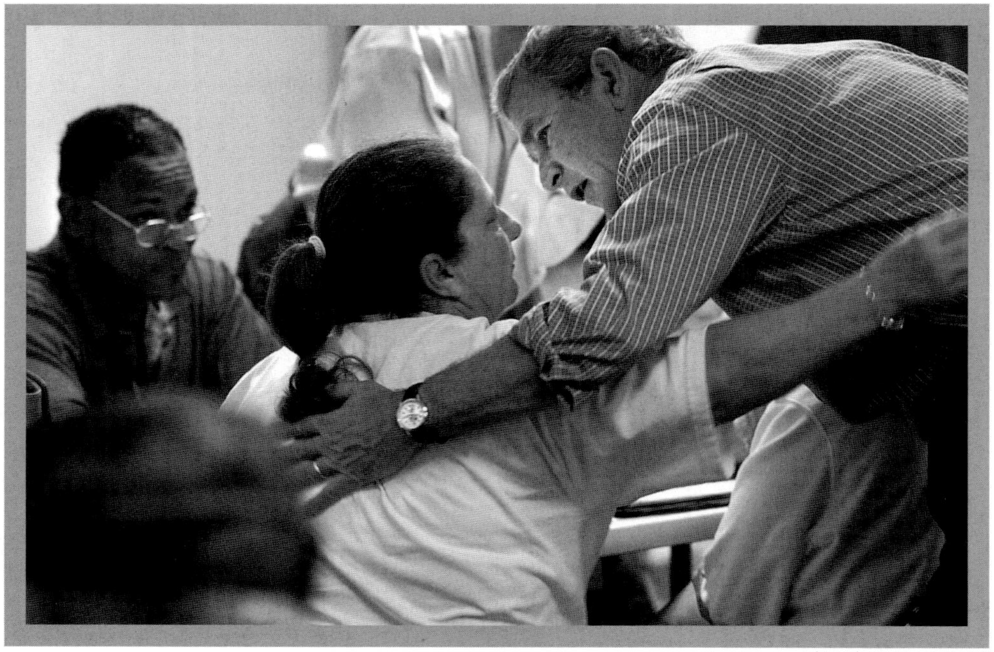

image by JASON NUTTLE

President Bush talks with hurricane victim Debra Bowen, left, of Port St. Lucie, during a visit to the Federal Emergency Management Agency relief disaster center at Witham Field in Stuart.

IN THE PATH

image by ALISON SIDLO

Marina Farias, left, and her son Eric Farias, both of Port St. Lucie, refasten a tarp that was covering their damaged roof. Like many Treasure Coast residents, the Fariases' roof was heavily damaged from Hurricane Frances and sustained further damage from Hurricane Jeanne.

In The Path

image by MOLLY BARTELS

Ty Strath reinforces what's left of his roof as the hurricane nears Vero Beach. Strath, who lives only a block from the Atlantic Ocean, sustained serious roof damage during Hurricane Frances. Strath and his family evacuated for Frances, but decided to ride out Hurricane Jeanne in their own home.

IN THE PATH

image by MATTHEW RATAJCZAK

The inside of Lorri Crouch's condominium at Atlantis III By the Sea on Hutchinson Island was destroyed when the hurricane pounded the barrier island. Sand, water, sea grapes and other debris broke through her hurricane shutters, filling her ground-level home, pushing her furniture and other possessions out the front door and into the parking lot. "The ocean took it away and your friends are going to help you rebuild," Crouch's friend Christy DeNoble said as they looked over the damage.

"The ocean took it away and your friends are going to help you rebuild."

In The Path

image by MATTHEW RATAJCZAK

"All I ever wanted was a small place on the beach to go collect shells."

Lorri Crouch, right, is comforted by her friend Christy DeNoble outside her first-floor condominium at Atlantis III By the Sea on Hutchinson Island in St. Lucie County after viewing her home, which was destroyed by the hurricane. In addition to the damage sustained from both storms, Crouch is also dealing with the Sept. 15, 2004, death of her husband, Dan. "It's not even so much the place," Crouch said with tears in her eyes. "It's the memories of things like us painting on the walls. Those are the things that hurt because I know I'll never have it back." The couple moved into their dream home on the beach three years ago. "My first inclination is to rebuild," she said. "But is this going to happen again five years from now? But it's the dream, I have to keep the dream alive. All I ever wanted was a small place on the beach to go collect shells. Dan gave it to me. So I'll rebuild."

IN THE PATH

image by SARAH GRILE

Phyllis Ewing, 56, who lives in Ranchland Mobile Home Park in Indian River County, reacts after finding out from neighbors Tom and Avalon McGann that her home has weathered the storm. Ewing had taken shelter at Oslo Middle School, staying there two nights.

In The Path

image by MOLLY BARTELS

Anne Lapardo, 90, of Barefoot Bay, assesses the damage to her home and property three days after the hurricane ravaged the Treasure Coast. Lapardo will not be able to live in her home on Barefoot Circle.

IN THE PATH

image by CHRIS WEATHERLY

Luann Henderson sits at the edge of Shady Rest mobile home park, a manufactured home community in Sebastian where her mother-in-law's home was destroyed. Henderson plans to return to her Barefoot Bay home, which sustained minor damage from Hurricane Frances and more from Hurricane Jeanne.

IN THE PATH

image by ALISON SIDLO

In St. Lucie County, Megan Dupuis, 10, left, and Sarah Perez, 9, pick up pieces of carpet left along the curb after Lakewood Park residents cleaned up from Hurricane Frances, only to have to deal with flooded streets from Hurricane Jeanne.

In The Path

image by MATTHEW RATAJCZAK

A resident of The Island Club condominium on Hutchinson Island in St. Lucie County looks out from one of her damaged rooms and adjusts a curtain. The building sustained serious damage from both hurricanes.

IN THE PATH

image by CHRIS WEATHERLY

Standing on top of The Old Opera House Gallery, Victor Smith assesses damage to the roof of the historical building on U.S. 1 in Sebastian. "The structural integrity is here so we'll rebuild it," said Smith, who is the boyfriend of owner Lisanne Monier.

"The structural integrity is here so we'll rebuild it."

IN THE PATH

image by JASON NUTTLE

Paul Foster, an employee with the Florida Drawbridge Co., makes his way down a 20-foot section of the Jensen Beach Causeway that collapsed during the hurricane, rendering it impassable. Foster and a co-worker locked the drawbridge into the upright position to let marine traffic flow freely. With the old causeway bridge to Hutchinson Island unusable, the project managers of the new bridge worked to open it as soon as possible.

IN THE PATH

image by ALEX BOERNER

Matt Ratliff, of Vero Beach, uses a metal detector to search the beach in front of the Vero Beach Inn. The hotel, already damaged by Hurricane Frances, also was hit by Hurricane Jeanne. About a dozen people, mainly employees and their family members, rode out the second storm in the landmark building.

In The Path

image by SARAH GRILE

Darren Chambers, of Spirit County, Tenn., helps repair the tongue-and-groove roof of Twentieth Avenue Church of God in Vero Beach. After the church couldn't find anyone to help fix the roof, Chambers, along with others from Spirit County who belong to the Church of God, volunteered to come down and help.

IN THE PATH

image by MOLLY BARTELS

Devonte Lewis, 10, helps move debris to a large trash pile on Booker Street in Fellsmere. Devonte was helping his great-aunt, Florence Lockett, clear a path to her front door. Lockett's home was condemned, as were many other homes on the street in Indian River County.

IN THE PATH

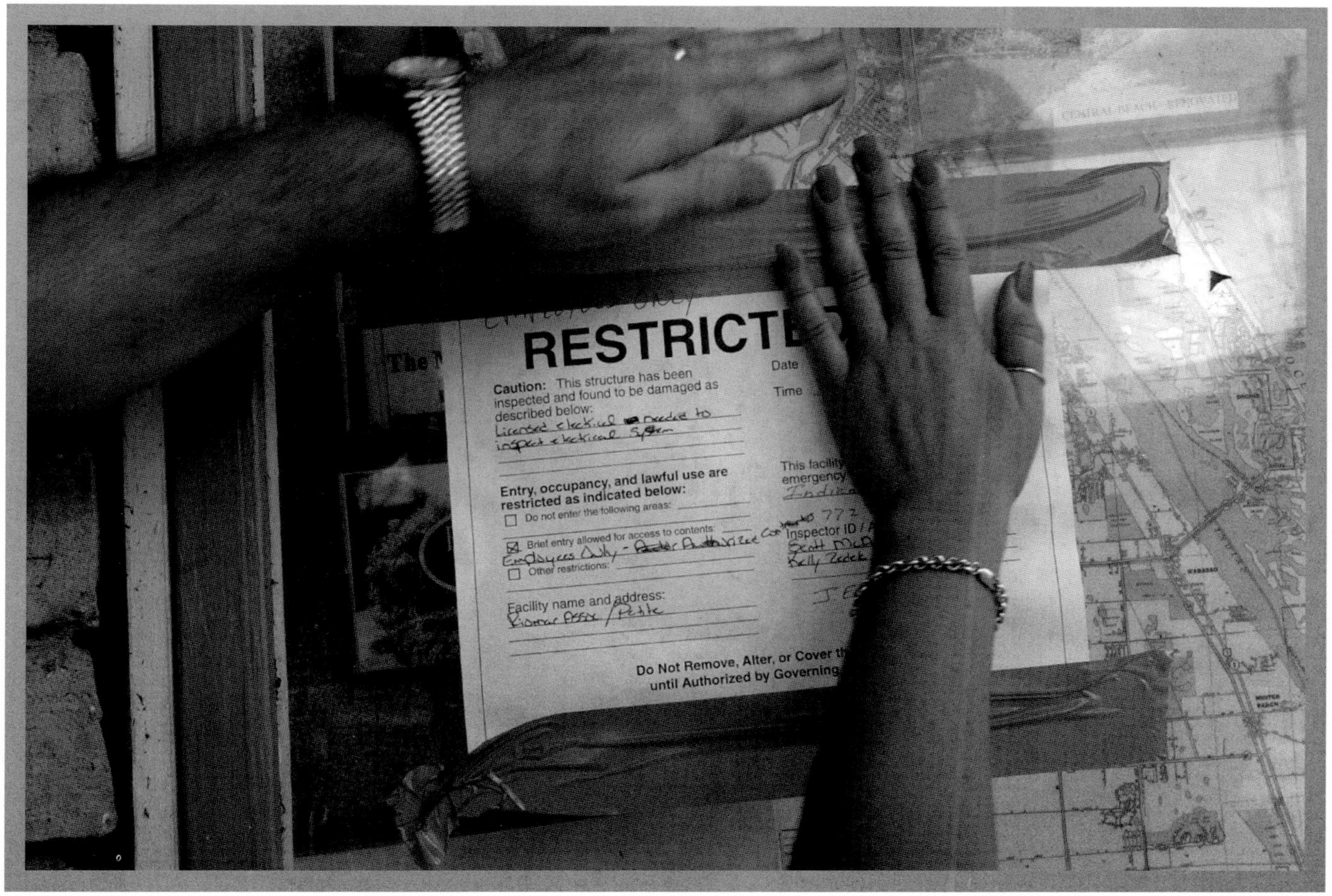

image by ALEX BOERNER

Kelly Zedek and Scott McAdam, with Indian River Code Enforcement, label a building as restricted while working on the barrier island in Vero Beach.

IN THE PATH

image by CHRIS WEATHERLY

Marilyn Couchlin checks out a beachfront home north of Wabasso Beach in Indian River County that was destroyed by the hurricane. Couchlin lives on the west side of A1A and received very little damage to her home.

In The Path

image by ERIC HASERT

Stephen Horne walks past his power generator in his yard on Hollyhock Drive Tuesday that caused his house to catch fire after the generator oil plug came lose Monday afternoon in Port St. Lucie.

In The Path

image by SARAH GRILE

Michael Ickes, who lives with his mother, Laura Seymour, on 4th Avenue in the Rockridge subdivision in Vero Beach, cleans his nails while sitting in front of his home. Seymour had just been able to discuss the damage of Hurricane Frances the day Hurricane Jeanne struck land.

IN THE PATH

image by ERIC HASERT

Rob Friedrich lights up charcoal and wood chips to smoke some ribs for a family dinner as his visiting friend Nicki Thompson sits inside his garage on College Park Road in Port St. Lucie. Friedrich used a generator while waiting for power to be restored to his home.

IN THE PATH

image by ERIC HASERT

Volunteer Diane Cuomo, of Port St. Lucie, cools off for a moment while helping hand out ice to residents lined up for supplies at the Federal Emergency Management Agency distribution site at City Center in Port St. Lucie.

IN THE PATH

image by MOLLY BARTELS

Andy Blossom, 18, of Sebastian, naps on his car while waiting for the power to be turned on at the Chevron gas station at 2602 U.S. 1 in Vero Beach. Customers waited several hours in hopes the power would return. Manager George Deep said the station had gas for 600 to 700 cars but had no way to pump it out. Blossom was one of the first ones in line and didn't plan on leaving. "I'm runnin' on fumes so I can't go anywhere," he said.

"I'm runnin' on fumes so I can't go anywhere."

IN THE PATH

image by MOLLY BARTELS

Sgt. Terry Tims, of the 1st 265th ADA National Guard Unit from Palatka, tosses a bag of ice to Pfc. Tracy Davis to load into a waiting car in the parking lot of Concordia Lutheran Church in Barefoot Bay. Members of the National Guard were assisting the Brevard County Fire Rescue Department in handing out ice and water to anyone who came there.

IN THE PATH

image by SARAH GRILE

Jim Davis, Indian River County Public Works director, center, speaks to the Rockridge neighborhood in Vero Beach about how the subdivision is in a flood plain and that is why the flooding was high during the hurricanes. Community residents gathered outside the clubhouse to listen to county officials and voice concerns in hopes of having questions answered.

In The Path

image by JASON NUTTLE

Helene Zothner, 95, of Palm City, waits outside the Federal Emergency Management Agency relief disaster center at Witham Field in Stuart to apply for assistance to fix damage to her home. Zothner is staying with her friend Linda Stitely until her house can be inhabited again.

In The Path

image by CHRIS WEATHERLY

Shelly Smith-Lamison and Nathan Lamison, lower center, grill their dinner for the evening in the parking lot of the Vero Beach Resort on State Road 60 as they visit with fellow crew members Josh Fry, upper left, Jeremy Harris and Bobby Carboni after work. Because of a curfew, contractors were left with little time to go out to eat. The crew members are subcontractors for Comcast.

IN THE PATH

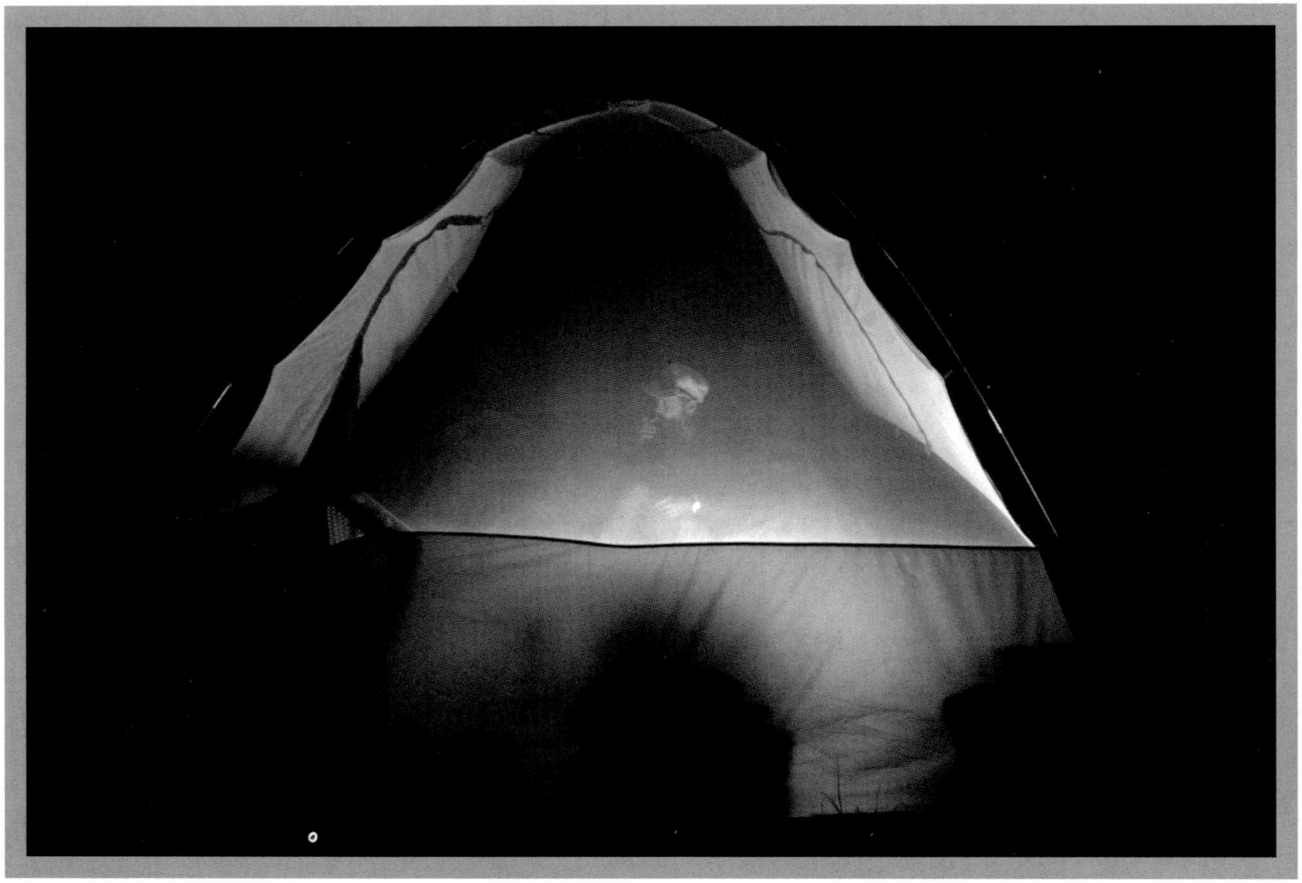

image by MOLLY BARTELS

Richard Sparling smokes the last of his cigarette before going to sleep in a tent outside his condemned manufactured home in Micco, just outside Barefoot Bay in Brevard County. Sparling and his girlfriend, Dawn Carroll (not shown) sleep in the tent on their rented property because they fear their landlord will consider the lot abandoned and seize the manufactured home, which Sparling and Carroll own. The couple's children are staying with family.

IN THE PATH

image by ALEX BOERNER

Rick Evans, left, of North Carolina, and John Daniels, of Virginia, unload aluminum at Mr. Scrap in Gifford in Indian River County. The men initially came down to work on debris removal from Hurricane Frances. They ended up staying in a hotel during Hurricane Jeanne, and went out collecting aluminum afterward.

IN THE PATH

image by JASON NUTTLE

Cathy Klopfenstein, office manager of Stuart Fine Foods, works on payroll for the store's employees from atop a cardboard box positioned near the store's front windows for light because the store is without power.

In The Path

image by MOLLY BARTELS

Ruth McPhee sits with her grandson's toy gun on the porch of friend Florence Lockett's home on Booker Street in Fellsmere. Lockett's home and several others on the street are unsafe to live in. Many of the neighbors gather on each other's porches to commiserate and help clean up the mess. McPhee's home on 100th Lane in Fellsmere also sustained major damage.

IN THE PATH

image by MOLLY BARTELS

Lauren Breinlinger, 12, takes a rest from helping her 90-year-old great-grandmother, Anne Lapardo, move out of her condemned manufactured home in Barefoot Bay in Brevard County. Lauren and many of her family members had been working all day removing furniture and family heirlooms.

In The Path

image by MATTHEW RATAJCZAK

From left, Oscar Borrientos, Oliber Escalante and Israel Perez untie Florida live oak trees at Calusa Creek Tree Farm in Stuart in preparation for the hurricane. The trees are untied so they can be laid on their sides or allowed to blow over in strong winds instead of snapping. The farm, still recovering from damage from Hurricane Frances, has been trying to meet the high demand for trees across the Treasure Coast and get the farm back up to 100 percent.

IN THE PATH

image by MOLLY BARTELS

Manuel Hurtado, right, and Jesse Hector Garcia move fruit packing boxes out of floodwaters near Riverfront Groves packinghouse in Vero Beach. The boxes had been secured to truck beds, but the hurricane's winds ripped them loose, scattering them in the area around Old Dixie Highway.

IN THE PATH

image by DEBORAH SILVER

Workers and residents sit behind the rail of the boat and relax during the approximately 30-minute ride from Sandsprit Park in Port Salerno to Sailfish Point. A ferry is the only way to get to the town after MacArthur Boulevard was washed out.

In The Path

image by DEBORAH SILVER

Sailfish Point homeowners and construction workers file out onto the private development's dock from the Safari 1, a fishing boat that Martin County chartered as a ferry after the only road leading to Sailfish Point was washed out. The free ferry runs every two hours, leaving from Sandsprit Park in Port Salerno starting at 7 a.m. The last boat leaves the park at 5 p.m., allowing workers on Sailfish Point to work until 6 p.m. "I still get a full day of work in," said Kase Breisch, a carpenter from Palm City.

IN THE PATH

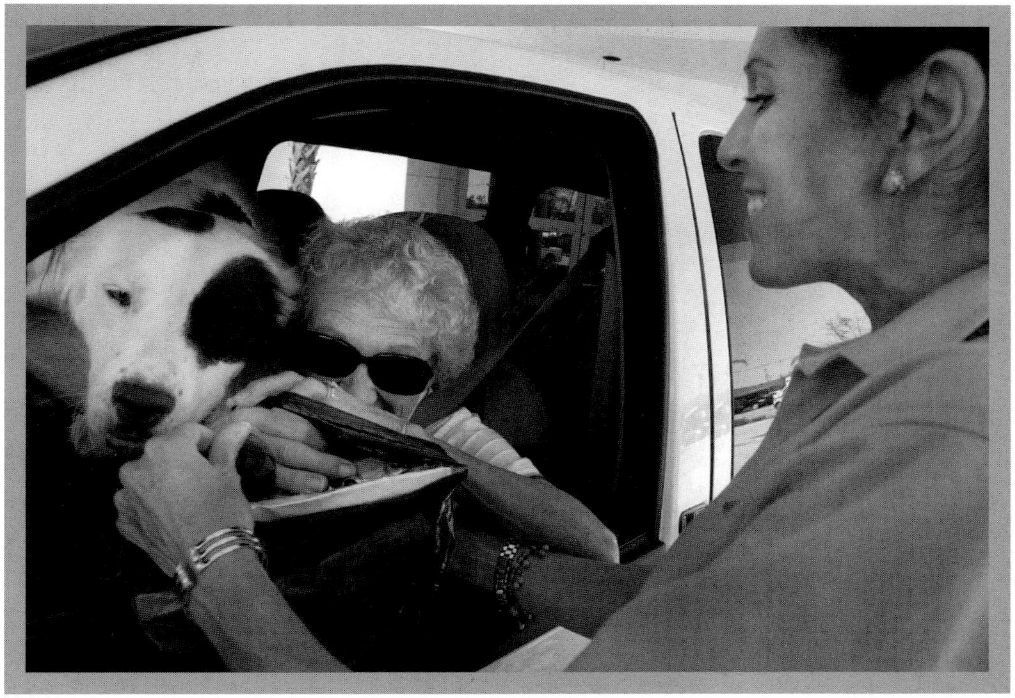

image by ALEX BOERNER

Sara, a border collie mix, smothers her owner, Doris Strong, of Sebastian, as she tries to get to a plate of food being handed into the car by Lynn Musselman, director of rehabilitation at the Sebastian River Medical Center. Musselman and other medical center directors were providing discount lunches to drive-through and sit-down customers as part of the Dine-In For Disaster program. Money made from sales of lunches, which were $5 each, went into a relief fund to be distributed to people throughout Indian River County.

In The Path

Fabiaro Sartos, left, and Josefa Silva carry trash past the figure of a great white shark that used to hang in the atrium of the education building at Harbor Branch Oceanographic Institution in Fort Pierce. Many of the research labs and buildings on the campus were damaged in both hurricanes.

image by ALISON SIDLO

In The Path

image by MOLLY BARTELS

Not fully understanding her family's grave situation, Natalie Strickland, 2, plays in the raindrops that fall from a leak in the ceiling of her family's manufactured home in Micco. That room and several others were destroyed in the hurricane. Natalie's family and many others who live on Fleming-Grant Road in Brevard County have received eviction notices from their landlord, Glenn Houston, after the hurricanes.

IN THE PATH

image by ERIC HASERT

Christopher Baldwin, 7, of Port St. Lucie, reads his story about his hurricane experience to his second-grade class at Manatee Elementary School in Port St. Lucie. It was the first day back in school for many St. Lucie County students after almost a month off because of the two hurricanes.

IN THE PATH

image by MATTHEW RATAJCZAK

A downed palm tree is reflected in the window of the learning center at The Boys & Girls Club of Martin County, Hobe Sound unit. SeaWind Elementary School kindergartner Morgan Ochwat, 6, of Hobe Sound, spends the day with 49 other children unable to attend their schools while they remain closed because of the hurricane. The club continued to provide relief for parents and kept the kids occupied until schools re-opened.

In The Path

image by JASON NUTTLE

Without power at home, Makayla Wilson, 4, left, and Ninosnka Estrada, 7, try to share a dry shirt after playing in a hose to keep cool in the 90-degree afternoon temperatures in Stuart.

IN THE PATH

image by SARAH GRILE

Marie Muller, standing in the doorway of her Heritage Plantation manufactured home in Indian River County, gives John Manfredi a knife as he prepares to put a tarp over her roof after the hurricane. Manfredi has known Muller for the four years he has lived in the park and spent most of his free time helping her and other neighbors and friends repair their homes and lives after the storms.

image by SARAH GRILE

John Manfredi rests while waiting for friend Cathy Melbourne to bring him a glass of tea while visiting her in Heritage Plantation in Indian River County. Melbourne, who recently bought a home in the park, discovered her home could not be lived in after the hurricane, so was temporarily staying at Kay Hibbard's, a friend in Vermont who also owns a trailer in the park. Manfredi is storing 17 plastic containers of Melbourne's belongings on his front porch. He says, "I've got room, they (people with damaged homes) can put it in it."

IN THE PATH

image by ALISON SIDLO

Rachel Chappell, 14, cools off by dumping cold bottled water onto her head while helping clean up a neighbor's yard in Martin County.

image by ALISON SIDLO

From left, Jessi Randell, 15, Miranda Spinweber, 14, and Rachel Chappel, 14, help clean up tree branches at John Richardson's home in Martin County. Teenagers with the Martin Grade Rustlers 4-H Club volunteered their time to clean up their neighbors' yards.

IN THE PATH

image by ALEX BOERNER

Jordon Fahmie, 19, of Sebastian, climbs up the storm-eroded embankment at Wabasso Beach in Indian River County. "We're checking out the stomping grounds — or what's left of it," said Fahmie, who was walking on the beach with his friend Matt Holderfield, 23, of Vero Beach.

IN THE PATH

image by ALISON SIDLO

Armando Aguilar, of SPS Contracting, bulldozes sand to build up the dunes in front of Island Dunes condominiums on Hutchinson Island in St. Lucie County. Crews were working throughout the area to rebuild vital sand dunes washed away during the hurricanes.

IN THE PATH

image by ERIK LUNSFORD

State Farm Insurance Adjuster Todd Douglas inspects damage to the roof of a home in Stuart. Douglas, who lives in Texas, moved temporarily to Florida with his wife and children to work on local claims for State Farm Insurance Co.'s catastrophe team. Douglas and other adjusters, including independent contractors, put in 14-hour days at the cramped makeshift office near the former Scotty's in Stuart. He said State Farm was well ahead of the curve for processing claims, which by his estimate topped 5,000 locally just for his company.

IN THE PATH

image by MATTHEW RATAJCZAK

Plastic tarps used to patch damage from Hurricane Frances, like this one in Jensen Beach, are now scattered everywhere, tangled in trees and power lines after Hurricane Jeanne blew through the Treasure Coast.

In The Path

A box truck parked between two cinder block walls lays smashed at Steve and Jerry's Furniture in Stuart. Hurricane winds caused severe structural damage throughout the Treasure Coast.

image by ALISON SIDLO

IN THE PATH

image by ERIC HASERT

Debris was found in unlikely places in Port St. Lucie.

IN THE PATH

image by ALEX BOERNER

Marjorie Hale, 25, of Sebastian, looks for shells along Wabasso Beach in Indian River County, where the only remnant of the convenience station left standing was a toilet.

IN THE PATH

image by MOLLY BARTELS

A couch that once was roadside debris left over from Hurricane Frances sits in the middle of the road in the Rockridge neighborhood in Vero Beach after Hurricane Jeanne struck.

IN THE PATH

image by MATTHEW RATAJCZAK

A swimming pool at Island Dunes Country Club on Hutchinson Island in St. Lucie County was turned into a giant sandbox with ladders going nowhere. The hurricane pushed sand into first-floor rooms, parking lots and other areas on the barrier island.

IN THE PATH

image by ALEX BOERNER

Kathy Parker, of Vero Beach, fights a gust of wind while standing behind The Ocean Grill in Sexton Plaza. Parker rode out Hurricane Jeanne in her home on the barrier island of Vero Beach.

IN THE PATH

ACKNOWLEDGMENTS

Photo Editors	Lloyd Young, Scripps Treasure Coast Newspapers Kelly Rogers, Scripps Treasure Coast Newspapers
Contributing Photographers	Molly Bartels Alex Boerner Juan Dale Brown Sarah Grile Eric Hasert Erik Lunsford Jason Nuttle Matthew Ratajczak Alison Sidlo Deborah Silver Chris Weatherly
Text Editor:	Lakara Webb
Design	David Melloch, Scripps Treasure Coast Newspapers
Printer	Nebraska Printing, Tampa

Front Cover

image by MATTHEW RATAJCZAK

Ricky Gould, 14, of Jensen Beach, sits on the curb next to what used to be the Jensen Beach Community Center on Sept. 26. Hurricane Jeanne destroyed the building that also sustained damage from Frances just weeks ago. "I took karate lessons in it," Gould said about the multi-purpose center. "I can't believe it actually fell down, it's been standing here for so long."

Back Cover - Before and after

images by ALISON SIDLO

(Far Left) Treasure Coast residents watch the incoming squalls from Hurricane Frances over the Indian River Lagoon in Fort Pierce on Sept. 3. The slow moving hurricane made landfall later that night.

(Near Left) After taking a direct hit from Hurricane Frances, the seawall at the same location along Melody Lane is in pieces.

*SING TO THE TUNE OF "LET IT SNOW"
12/04
OHH HURRICANE SEASON WAS FRIGHT-FUL...
...DUNES VANN-ISHED BYY THE BITE-FUL...
jparker@flatoday.net
...WE WERE LEFT WITH NO PLACE TO STAND...
...LET IT SAND LET IT SAND LET IT SAND
From Uncle Sam with Love~
caglecartoons.com
JEFF PARKER ©2004 FLORIDA TODAY